Climate Change

Climate Science Facts & Fiction, & Tackling Global Warming

Anna Revell

Copyright © 2017.

Table of Contents

Introduction

Climate change is here. And it looks as though it is going to stay.

We are experiencing its effects already, but the evidence suggests that matters will only get worse.

Those following the news this year, in 2017, will be familiar with the impact of hurricanes on the northern West Indies.

Nations have been destroyed, people killed as landfall was made by successive storms in a short period. The small island of Barbuda in particular suffered devastation.

There was a very significant satellite image of the region reproduced in numerous newspapers.

It showed a line of three hurricanes, the first having passed the West Indies, the second over it, north of Cuba, and the third about to hit.

Climate change is already a serious threat to the lives of mankind and, indeed as we will see later in this book, every other living occupant of the planet.

Your reading will offer an insight into the causes and effects of climate change. It will provide case studies alongside practical guides as to how we can all help.

It will also look at the climate change deniers, that group of often powerful people who fly in the face of accepted research in an attempt to justify their continued use of fossil fuel. They frequently seek justification for their pumping of greenhouse gases into the atmosphere.

If, like many people, you are already concerned by the risks of climate change, this book will provide hard evidence to support your worries.

If you are a denier, it will aim to present enough information to make you question your views.

Wherever you stand, your understanding of this major and growing problem will be clearer.

Chapter One - New Orleans

Not so very long ago, in August 2005, New Orleans was devastated by a natural phenomenon.

Hurricane Katrina changed our views of these storms, because it hit a major American center of population, causing damage and loss of life on a scale rarely encountered on mainland USA.

While it is not right, it is perhaps inevitable that when natural disasters hit elsewhere, we shake our heads, offer sympathy and, perhaps, support. Then move on with our lives.

But if they hit us directly, our views are different. They are stronger, and we expect more action to be taken by the leaders of our nations.

Hurricane Katrina began its journey in the tropics, as a depression. It turned north, just missing the Bahamas, and by this time it was powerful enough to be categorized as a hurricane.

Next, it veered west, just catching the southern tip of Florida. It spent only six hours over land, at this stage a category three storm.

Then, fueled by the warm waters of the Gulf of Mexico, it strengthened, veered sharply north and, at 10.00am on August 29th made landfall just to the south of the Louisiana city.

By now it was a major, category five, storm.

The wind speeds within the hurricane were running at a sustained velocity of 200km per hour.

Alabama and Mississippi were also badly affected, with Biloxi and Mobile two cities to suffer.

Florida, too, experienced sustained black outs and damage to homes and businesses, but it was New Orleans that suffered most.

Here we saw what can happen, even in the developed world, when nature does its worst.

The fuel of hurricanes is the seas over which they travel.

When hurricanes make landfall, they lost that fuel, and their energy is reduced.

Crossing the coast saw Katrina weaken rapidly, deteriorating from its full scale, category five, high to just – if that is the word – category one as it headed inland.

Soon it was downgraded from hurricane status to that of a tropical storm. Although huge in size – satellite images show it filling almost the whole of the Gulf of Mexico – its winds were not as destructive as initially feared.

But what was worse than expected was its storm surge.

This was twenty-eight feet high at its extreme and in New Orleans itself almost twenty feet. Added to this, especially on its western edges, up to ten inches of rain fell in a short time.

Three days before the hurricane hit the coast; both Louisiana and Mississippi instigated their emergency procedures.

From New Orleans, around 1.2 million people were evacuated, but that still left 100000 who, for reasons such as medical problems or lack of transport, could not get away.

About 10000 of these headed to the Superdome to see out the storm. The remainder hunkered down to ride out the weather.

New Orleans is braced for hurricanes. Most of the city lies below sea level, and pumping stations are there to clear the city of rainwater if needed.

A complex series of levees protects the population from flooding.

But Katrina was huge, and its storm surge even greater. Coastal structures were destroyed when land fall was made, and the pumping stations were quickly put out of action.

Then, the levees were breached, with no less than fifty points at which the water escaped, pouring into the City. New Orleans was flooded completely, up-to fifteen feet in some places.

Those who had not escaped the city were overwhelmed by the speed of the surging sea. Many scrambled onto the rooves of their houses as the water levels continued to rise.

Meanwhile, in the Superdome, conditions in the hot and humid city were becoming unbearable.

Violence and widespread looting began to occur and, for a time, law and order was lost.

Of course, it was the poorest that had been left behind, and these people were mostly black African Americans.

The reaction of the Federal Government was inadequate in the face of such suffering, with President George W Bush at his most lost and vague as he struggled to voice the feelings of his people, and provide help.

Those trapped in the Superdome were left without any support, although it had been promised, and the police desperately pleaded for anybody with a boat to help rescue the thousands who were trapped elsewhere in New Orleans.

At its worst point, eighty per cent of the city was under water.

Meanwhile, white politicians congratulated each other on their work.

The social and racial divide could not have been greater. Over 1800 people died, and ten years after Katrina hit, the population of New Orleans had crept back to only 80% of its previous figure.

Each of us who remembers Katrina, recalls how destructively serious the natural disaster was. How it was exacerbated by the inadequate human maintenance of levees.

How the Bush regime did so little – the President flying over the city and not even getting his feet wet. How many city officials simply lied to keep their jobs and reputations safe while the dead lay drowned on the streets?

Billions of dollars were, apparently, pushed into getting the flood defenses up to scratch in the increasingly likely event of another major hurricane hitting the city.

But nobody who lived through the carnage, or who lost relatives or their homes and possessions, really holds complete faith in the future.

If another category three storm struck, who would bet on the city withstanding the impact? Not many.

But why do we start this book with a recount of one of the worst natural disasters in US history?

The answer is simple. It is that if climate change is not addressed, quickly, that fate that befell the citizens of New Orleans will become an ever more likely occurrence.

Chapter Two - Climate Change, Some Definitions

Let us start this chapter with a few definitions, since there is often confusion about the precise meaning of terms around the subject of climate change.

We will look at weather, climate, global warming and climate change, since these are all different, but interlinked, phenomena.

Weather

Weather is the immediate atmospheric conditions experienced over the short term, such as from the immediate, to a couple of days ahead.

So, rain, snow, wind, storms and so on are examples of weather. Even the rain and destructive winds of Hurricane Katrina were examples of weather – particularly extreme weather - but nevertheless, it was such.

Climate

Climate, on the other hand, is much longer term. It means the regional – even global - averages of weather that are experienced. So, it is the average humidity, rainfall and temperature of an area.

Micro Climate

A micro climate is a climate that affects a small area when most of its surroundings experience different conditions.

So, for example, the Isles of Scilly are a group of islands just under 30 miles off the south west coast of England.

However, they lie in the Gulf Stream, a current of warm water that heads up from the tropics. Visit Scilly and you will see tropical plants in abundance, which is completely different from the usual flora of the larger region.

This is because the micro climate of Scilly is warmer, and hence the plants can flourish. Indeed, it is not unusual to watch the British weather forecast and see a day of cloud and rain predicted.

Then, to sit on a beach on Scilly and see perfect blue about and around, with heavy clouds which never get closer hanging in the distance.

Climate Change

Climate change is the slow but persistent change in the world's climate. It is believed that this has largely been caused by the burning of fossil fuels, such as coal and oil.

A name is given to this man made destruction – it is called anthropogenic climate change. Some deniers of climate change accept that the world's temperature is rising.

However, they do not feel that it is the fault of man. Rather, they see it as a part of a natural cycle of climatic change that has always been a feature of life on earth.

However, most scientists agree that climate change is anthropogenic, and it results from the greenhouse effect.

Here, the gases released trap heat inside the earth's atmosphere. As a consequence of this, the earth heats up and climate change effects include the melting of the ice caps at the north and south Polar Regions.

This has led to rising sea levels, and an overall rise in global temperatures. An outcome of this is that there are increased cases of extreme weather.

Such as the increase in intensity of tropical storms that we discussed in the introduction and last chapter.

But it is not just these extreme weather examples we are witnessing.

Indeed, watch the weather forecast regularly and it seems that not many weeks pass without some superlative ascribed to the weather. 'Worst storm, hottest month, wettest week' and so on.

Global Warming

Global warming is one aspect of climate change. It is thought by many to be the route cause. It describes the gradual increase of temperatures on Earth since the beginning of the last century.

In particular, the burst in global warming from which we are now suffering accelerated during the 1970s. This was a time of rapid expansion in world industry. The fossil fuels burned as a result of this upsurge account for the increase in warmth.

If we are to lay the blame for the increased numbers of extreme weather, and the general rise in sea levels and melting of the ice caps on global warming, then it is important to understand how this occurs.

How Global Warming Occurs

Global warming deniers make the quite valid point that fluctuation in the Earth's temperature is normal.

After all, the Ice Ages of the past occurred when, as far as we know – alien life apart! – There was no burning of fossil fuels.

However, the rate at which the earth is heating up, and the prolonged period over which this has occurred, combined with what we know about how greenhouse gases work means that it is hard to take climate change deniers seriously.

Any average temperature increase of a degree Celsius over a century or so would be considered as evidence of global warming.

And currently, we are far exceeding this rate of warming.

Simply looking at graphs plotting changes in the northern hemisphere in the latter half of last century paints a very clear picture.

Mean temperature shows an increasingly rapid rise, sea levels go up even more sharply while snow cover exhibits a clear downward trend.

But when we are talking about climate change being a phenomenon that occurs over time that is very much what is meant. It is a very long-term procedure.

So, whilst it is undeniable that there are notable changes in climate, the deniers would argue that this does not represent climate change. This warming has not gone on, they say, for long enough to be significant.

Even a century is only a short period when we talk about climate change. So, if the burning of fossil fuels is seen as a cause of global warming, and the industrial revolution only began, globally on a very small scale, two hundred years ago, then there has not been a long enough period to really judge its impact.

We might reply that they should make this comment to the citizens of New Orleans, or the Caribbean.

But, the rapid increase over the last fifty years, nearly half a degree Celsius, would seem to counter this argument.

And that, in a nutshell, is one of the big problems with global warming. It needs to happen for a long enough period to be registered as a concern.

But by the time sufficient decades have passed, the damage done will be too great to repair. The effects of pumping tons of gases into the atmosphere will not diminish just because we stop doing so.

Those effects will take many centuries to erase.

We can still address climate change, and really need to do so because there will come a point when the damage caused is too great. That point is getting ever closer.

But, we need to wait until it is too late in order to prove to the deniers that it is a serious, catastrophic problem.

Were these deniers simply oddballs living in their own worlds, it would be a small problem. But when world leaders – President Trump is currently the most obvious example – realize that they can gain economic advantage by saying that climate change does not exist, and then the world has a problem.

And that is because such leaders will exploit their advantage by using cheap, fossil fuels (while they still exist) while morally stronger leaders invest in non-polluting energy sources. Economically, because financial costs are higher, the ethical energy users lose out to the exploiters.

It is, of course, ever thus. Short term gains against long term disaster.

Hard Evidence

So, let us consider more of the evidence supporting the fact that global climate change is real. Records on global temperature began to be kept from the 1800s.

1944 showed a sudden and high peak in temperature – it was the hottest year by far since records began.

But, in the last twenty years, only one (1993) has shown an average temperature lower than this high.

Next, the ocean's temperature is increasing in a very worrying way. Since the oceans drive our weather, any changes are a cause of concern.

Temperature increases are not just on the surface, but down to an incredible 3000 meters increases are apparent.

If global temperatures are a cause for consternation, then what we are witnessing at the poles are even more worrying. Their increase in temperature is double the average. As a result, and as might be expected, snow and glacier coverage is receding in these areas.

Rainfall patterns are changing. Parts of Europe, Asia and the Americas are seeing much increased rainfall, while, for example, the Mediterranean, and parts of Africa are experiencing long dry periods.

This has led to intensive and long-lasting droughts.

It is also becoming windier. Westerly low pressure systems are deepening with greater regularity, and the result of this is stronger winds, allied to heavier rain.

Extreme temperature types are changing. There is less in the way of cold nights, and freezing days, but heat waves are much more common.

Finally, while there does not seem to be an increase in the numbers of tropical storms – something the climate change sceptics are keen to tell us –their intensity is much greater.

Something the same people are less keen to share.

Anecdotally, most of us would know that that it is wetter, or drier, than usual. That there is, for many, less snow and ice.

But such evidence is more than anecdotal. It is scientific fact.

How is Global Warming Caused?

Global warming is a result of the greenhouse effect increasing.

The greenhouse effect is essential for most life on earth to exist. Without it, we wouldn't be here, because the world would be too cold to live on.

However, it is a balancing act. Too much greenhouse effect and we have problems.

What is the Greenhouse Effect?

You can think of the greenhouse effect through the analogy of a car.

If the sun is shining (it always is up in the atmosphere where the greenhouse effect lives) and you leave your car in it, then when you return it is warmer inside the vehicle than outside in the fresh air.

This is because heat that has gone in to your car has been absorbed by the seats, the dashboard, the fluffy dice etc.

But when that heat is released, the windows stop some of it escaping. So, more energy (in the form of heat) has gone in than can get out.

That is what is happening on earth. And the window effect that occurs in a car is replaced by the greenhouse gases released when fossil fuels are burned.

The heat from the sun can pass through them one way, but cannot escape the other. We are all basically living in a car parked in the open on a hot day.

More specifically, NASA estimates that about 70 per cent of the sun's energy, in the form of heat, reaches the earth's surface, where it is absorbed by the ocean and land.

The other thirty per cent is reflected back off of clouds, glaciers and other reflective surfaces.

The heat absorbed by the earth is radiated out and some of this goes back into space. However, gases such as methane and carbon dioxide absorb some of this energy, ensuring that the earth stays warm enough for life.

It is when there is too much of these gases, such as caused when fossil fuels are burned, that too much heat is trapped, and the earth begins to warm.

That, most scientists believe, is the problem we are experiencing.

But the change is self-perpetuating. We said earlier that the giant ice fields are one of the features of the earth that reflect the sun's heat back into space.

As this ice melts, so the amount of material to reflect heat back is reduced, meaning even more heat is absorbed.

Chapter Three – Two Case Studies from Africa

When Hurricane Katrina hit the US, it caused widespread damage, death and suffering, as we saw in the opening chapter.

But, however poor the Federal Government's response, however distressing the self-serving reaction of some local government officers, and however unjust that it was the poorest sections of society that suffered most, the US is, nevertheless a hugely powerful developed economy.

But many of the suffers of the worst impacts of climate change are not just the poorest in a rich country; they are the most deprived in the poorest of countries.

Ethiopian Coffee Farmers

Ethiopia is a case in point. Let us look at a specific problem which climate change is causing. About 15 million Ethiopians survive by producing coffee beans.

Their livelihood is under severe threat. The country is especially prone to drought, and it remains politically unstable in many regions.

The borders around the Somalian region are extremely volatile, terrorism is rife and the border with Eritrea is closed.

The coffee trade is essential to the country's economy; it provides a quarter of its export income.

The Arabica bean lies at the heart of the problem. Its smooth taste appeals to Western palates. But the bean is vulnerable.

Experts from Kew Gardens predict that, within fifty years, it may have disappeared from the country.

Warmer temperatures are speeding up the ripening process, so flavors do not have the time to develop.

However, it is not just direct temperature issues that are threatening the bean production.

Global warming is leading to drought and growing deforestation. Pests', including a beetle called the coffee berry borer is particularly dangerous, able to devastate crops.

The beetle can be largely controlled naturally as it is prey to number of species of bird. But, the gradual loss of the birds' habitats is leading to falling numbers.

The beetles thrive, and spread and the coffee problem is made worse.

Another species of coffee bean, Robusta, is more resistant to attack and harsh conditions, but its flavor is bitterer, and there is little farming of it in the country.

The Western market is simply not there.

Without the Arabica coffee bean livelihoods will be lost and income will fall. The entire region is unstable, and recovering from a long lasting civil war.

Without income sources, poverty will grow even more, and the risk of further conflict increases enormously.

In order to have any realistic probability of the industry surviving, deforestation must be addressed, and the climate needs to deliver cooler temperatures.

The Western world might regret losing their morning coffee, but Ethiopians face losing their livelihoods.

And that could lead to disaster in the already troubled regions.

Wildlife in the Democratic Republic of Congo

It is not only the human race that is affected by climate change. In the central African country of The Democratic Republic of Congo, many species of animal life are under severe threat.

The Salonga National Park is Africa's largest tropical rainforest reserve. It is accessible only by water.

The region is already under threat through war, and the park is subject to the risks of deforestation as well as poaching.

It is the sole remaining habitat for many endangered species. The dwarf chimpanzee, the Congo peacock and forest Elephant all remain at sever risk of extinction.

Indeed, it is predicted that the damage caused by global warming will result in the loss of seven species of mammals by the middle of the century.

By three quarters of the way to the year 2100, another eleven are likely to disappear from earth forever.

It is, admittedly, not just climate change that is resulting from man's actions. Logging is increasing in the region, causing severe deforestation.

Migration into the region, as a result of increasing population, associated road building and agriculture are all adding to the problem as well, in the DRC as a whole.

But it is climate change that is the main problem. Remembering that an increase of a degree a century represents a disturbing and dangerous amount of global warming, in central Africa temperatures are rising by three times that amount.

At the same time, precipitation is dropping by 3 or 4 per cent a year.

Man's colonization of the region is also restricting natural migration of animals to cooler and moister regions.

Even were species able to enjoy freedom of movement, predictions suggest that about twenty per cent of species in Africa will be extinct, or critically in danger by the end of the century.

If they remain in the central African basin, then forty percent of species will, by the end of the century, be gone.

Within the lifetime of children born before the end of the decade, the risk must exist that mammalian animal life could be more than halved in Africa.

That is a very serious state of affairs, and places a huge responsibility on the West.

Surely, it is more important that a new petrol or diesel-powered car?

Chapter Four - Deniers' Myths

It could be for political or financial gain, it could be for genuine beliefs in the science that disputes climate change, or skepticism towards the substantial research supporting climate change.

But, whatever the reason, those who deny that climate change and global warming are a cause of real concerns, and they employ many myths to give credence to their arguments.

We will, in this chapter, consider some of them, and evaluate their merits.

A common claim is that there is no consensus on climate change, that it is the theory of a small group with vested interests who have somehow persuaded the rest of the world that there is a problem.

Well, there certainly is consensus, one which is growing steadily.

Ninety Five percent of scientists who are looking into climate science identify rises in temperature as being, in the main, caused by human activity.

In the twenty years between 1991 and 2011, undertaken by researcher John Cook, found 12000 publications and papers which supported that climate change was occurring.

There are opponents, who dispute the findings. Their language is often extreme, and not terribly scientific. For example, in an article in the Wall Street Journal, two members of a libertarian think tank argued that climate change was not occurring.

One of them was renowned for referring to his opponents as 'global warming Nazis'

Another argument used by deniers is that, while climate change might be occurring, this is just a natural part of the earth's climatic life.

As we saw in the beginning of the book, there could be some validity in this claim. Climate, by definition, has to be looked at over the long term.

But, we now know that climate change is a result of the greenhouse effect. We understand the gases that contribute to this effect.

We know that, since industrialization and especially since the 1970s, greenhouse gas emissions have gone through the roof, so to speak.

The difference today is that it is only man-made factors that are adding, in any significant way, to the increase in the greenhouse effect.

There have been no natural causes that have been identified as causing the change.

Some argue that it is, in fact, solar energy that is causing global warming. Heating, in other words, from outside the earth's atmosphere.

Again, no science supports this. The troposphere is warming, certainly, but above that the stratosphere is currently cooling.

One of the more bizarre myths occasionally employed is that climate change is there, but it is actually good for the planet.

It is extremely difficult to offer any basis for this argument. But then again, those who make such claims are often not too bothered by supplying evidence.

We would just ask those who perpetrate this view to consider droughts, famines, wild fires, floods and extreme weather conditions currently happening more and more often.

Or to consider the animal species dying out. An open-minded person would have to change their view.

Equally bizarre is the complete denial of global warming, with the statement that the planet is cooling.

It's quite difficult to argue against this idea. Just as it is very difficult to challenge a viewpoint such as the earth is flat or there was no life prior to 8000 BC.

If the evidence is just ignored or denied, there is not much any reasonable person can do.

Some deniers make the point that evidence suggest that there has been no warming for eighteen years.

It was quickly proved that this assertion was wrong, and that the planet is actually heating up ever more quickly.

The simple trick of measuring data accurately dispels the myth that global warming is over.

Climate change sceptics include a number who love a conspiracy. Climate change is just a theory to damage America, the West, is promoted by communists (or maybe, currently, North Korea?)

This kind of supporter was delighted by allegations of tampering with information by climate change scientists, in the 'Climate Gate' case.

In the UK, the House of Commons Science and Technology Committee exonerated the University of East Anglia from allegations of tampering with data.

At the same time, investigations in the US cleared anybody accused of wrong doing. They reported that:

'There exists no credible evidence that any scientist had or has ever engaged in, or participated in, directly or indirectly, any actions with intent to suppress or to falsify data.'

No sitting on the fence there.

Misinformation is often used by deniers, or evidence that is fragile in the extreme. Such claims might include statements such as the sun is warming up or the ice sheets are growing.

The sun, though, is on the whole cooling. Temperature fluctuates, but it is widely accepted in science that the temperature of the sun does not influence our planet's warmth.

Some deniers take the idea further, saying that Mars is also warming up; therefore the phenomenon cannot be man-made. But this is overlooks that the temperature of Mars is largely due to the planet's orbital changes and the angle of its tilt.

There is some growth in the ice sheets, but that is limited to a small area of Eastern Antarctic. Overall, there is a reduction in the amount of ice at the poles.

Climate change has never been linear. Whilst the trend is upwards, different places at different times experience local variations. That would be normal in something as organic and subject to so many factors as climate.

Those who oppose the theory of climate change do, as we have seen above, take a scattergun approach to finding an argument that can stick.

Sometimes they deny that the earth is warming, sometimes they say that it is fine that it is happening. At other times, they say it is a natural process, unaffected by man.

They claim that it is short term, and models are no more than science fiction, on which we should place little trust.

Losing competitive advantage in industry is not worth it, they say, when those in Kyoto may be wrong.

After all, it is not much of a gamble. Decimation of animal life, loss of entire tracts of human civilization and chronic natural disasters are a small price to pay for cheaper fuel.

Sorry, that may be a little cynical!

But one of their most alarming claims is that climate change is so far advanced that nothing can be done.

Such a defeatist argument is worrying. After all, if world leaders and industry chiefs feel that nothing they do will help to preserve the planet, then they will do nothing.

However, their arguments are false. History has proven how quickly the planet can recover. Reducing the amount of carbon emissions and protecting the landscape will make a difference, and quickly.

But, we need to move soon.

Chapter Five – Waves of Trouble

Two impacts of climate change have led to worrying rises in the levels of the world's oceans and seas.

Firstly, global warming has led to melting of the land based ice caps, and polar glaciers. Of less significance, but still worthy of note, the heating up of the sea has led to expansion in its size.

Put simply, when water heats up, its volume increases. We see this when we overfill a kettle. At cold temperatures the water stays inside the implement, but as it boils, it begins to bubble out of the spout.

This increase in sea levels is worrying enough itself, particularly given man's tendency to build on coastal regions and low-lying country.

The Climate Institution measured the sea levels in 2014, compared to an average over the ten years previous to this.

The results are conclusive. There are small areas of slight drops; the oceans near the poles are slightly lower, but only in places. There are narrow bands where the level has stayed constant.

But great swathes of the oceans show rises, some approaching 10 cm. Equatorial Pacific and the tropical areas of both the Atlantic and Pacific show substantial increases. The

Indian Ocean and Southern Pacific seem especially to have suffered in this respect.

Of even more concern is the trend that shows sea levels rising at an increasingly rapid rate.

Climate specialists are busily mapping out the consequences, identifying both the likely outcomes, and worse case scenarios.

It seems as though the rate of sea level rise is greater than it has been for three millennia. It has risen by almost three inches in the last twenty years.

The process is self-perpetuating. As we saw in the previous chapter, sheets reflect back heat, but as these diminish, so more heat is absorbed by the earth. The seas, especially, absorb large quantities of heat. And they are getting bigger.

Within the lifetime of many of today's children, by the end of the current century, the sea will be two meters higher than at present, according to predictions.

Even had the Paris accord been adopted and adhered to (i.e., a maximum temperature increase of two degrees), significant increase in sea levels would happen. But with many parts of the world ignoring obligations, the situation looks worse.

The upshot of all of this is that many coastal population hubs are taking actions to mitigate against the risk. As ever, it is those countries with the greatest wealth that can do the most.

In the US, several cities have begun to put into place plans to save themselves. For example, models suggest that, within thirty years, significant parts of the New York could be under water.

The process could start within the next decade. Staten Island, Brooklyn and the long arm of Queens look especially vulnerable.

But New York officials are undertaking enormous defense plans, involving the protection of natural barriers, such as sand dunes, and the laying of offshore breakwaters to keep waves associated with storms from endangering the city.

The process is complex, intrusive and very costly, but absolutely necessary if the city is to be saved from the effects of another major storm, such as the recent hit from the vestiges of Hurricane Sandy.

Further north, the city of Boston is also vulnerable. A storm surge could be a particular problem. This is not least because it would be something from which Boston rarely suffers.

The residents and city would be prepared, for example, heavy snow, but not the impacts of a hurricane.

However, Boston has taken a note of the events in New Orleans. Its plans place a priority on the particularly vulnerable.

It focuses on supporting low income households and small business owners. For example, in the need of evacuation from the city, buses would be directed to the poorest regions.

It is not just the US where preparations are underway to deal with the prospect of danger from rising sea levels.

In Australia, cities were built with the knowledge that storms and associated surges are regular occurrences. However, they were built under the expectations normal sea levels.

To take a particular example, in Sydney all sea walls were assessed it determine their effectiveness against the ocean. Many are old, and would probably not survive the combination of heavy storms and higher levels.

The sea walls are then strengthened, or raised, to cope with the danger. Whether or not this will be enough is, of course, open to debate.

It tends not to be storm surges and hurricanes that create the risk in Europe. The continent is not prone to them. However, tidal flooding is a problem, and there are surges created by heavy non-tropical storms.

And, in some parts of Europe cities, such as in the Netherlands are built below sea level.

Here, levees offer protection, but we saw in New Orleans that these do not always work.

A case was in Rotterdam in 1953, when the dikes – walls or embankments to protect a city from the sea – were breached and over 1800 people died.

The Dutch Government reacted by rebuilding the dikes, but in doing so impacted on aquatic life and the general environment.

Now, however, they are seeking to create more eco-friendly defenses.

The Climate Institute offers all cities and regions three pieces of advice which can help defend against invasion by the sea.

There are firstly to protect natural barriers, such as barrier islands. These have offered defense for centuries upon centuries, including times when sea levels have been high.

It is just that mankind has no idea of when sea levels will stop increasing.

Secondly, the man made barriers should be checked, strengthened and reinforced. But, climate change is such that it is feared these two measures will not be enough.

Their third suggestion is to educate people about the threat, and in their planning, ensure that the most vulnerable – the old, sick, young and poor – receive assistance.

Chapter Six – What Can Our Leaders Do About Climate Change?

In the autumn of 2017, the United Nations Secretary General Antonia Guterres addressed the issue of climate change, seeking to find ways to best prepared for the Climate Summit of 2019.

Guterres highlighted six areas which could be addressed. These are:

1. Investment in clean technology

2. Carbon pricing

3. Energy transition

4. Risk Mitigation

5. Supporting the contribution of sub-national actors and business

6. Mobilizing finance

In addition to these, several members also raised the importance of:

7. Deforestations

8. Ocean degradation

Let us look at these in turn, and see what they actually could mean for the planet, in the light of the 2015 summit on climate change in Paris, and in anticipation of the 2019 follow up.

Clean Technology

This is a broad term that refers to recycling, renewable energy, information technology, green transportation, biofuels and such like.

Let's take, to explore further, probably the biggest issue with clean technology, and that is the contest renewable energy faces with fossil fuel energy.

It is a bit like a person on a diet. They know that the salad and oily fish are what they should be consuming, but the burger is so easy, and tasty.

Although it is running out, there remains an awful lot of fossil fuel in the ground. And it is worth a lot of money. This usually talks.

Although nobody can say for sure, by the time all the fossil fuels in the world have been burned, maybe fifty years from now if renewables do not take up more of the supply, the world could have doubled the two degrees aimed for. Who knows what a four degree increase in temperature might do, and in many of our lifetimes?

The problem is that what has been done in the last ten years just hasn't made a difference. We might see farms of wind turbines, offshore wave energy systems and solar panels on top of country cottages.

But these moves are just tickling the sides of the problem. Much of the world, developing and developed, continues to pour carbon dioxide into the atmosphere.

Certainly, journeys of forty years ago in many developed countries, when separating the smoke of factories from clouds may be much less frequent. But the rest of the world is more than compensating for that.

In order to get the climate back to two degrees increase which gives some chance of the slowing the trend, if not actually turning it back, is not on course. For all the words of the UN, and well received as they are, the organization simply does not have the power to bring about the change it wants.

Carbon Pricing

Carbon pricing makes sense. Put a price on carbon use, and users will be forced to look at alternatives.

There are two main ways of imposing such a cost – through taxation and through a usage cap, called cap and trade. IN this second option, a Government sets a limit of use of carbon fuels and reduces this over time.

Again, an excellent way of controlling carbon use.

So, where is the problem? The problem is that the UN cannot enforce, it can only discuss and advise. When there are global economic uncertainties, Governments will place their own interests before that of the world.

It is sad, but true. Just ask Donald Trump.

Energy Transition

This is a term to explain the move from one energy form to another. For our planet, most energy is fossil based, and in order to address the gradual diminishing of sources and, more importantly, address the damage being done to the planet, many countries are seeking to move to renewable sources.

But, as we saw above, there are pressures that are stopping this from occurring as quickly as it might.

Unfortunately, so many of these moves as defined by the UN are dependent on Governments agreeing to support them.

There is evidence to show it works. Sweden is one of the greenest countries in the world when it comes to energy use.

But it is also one of the most competitive when it comes to trade deals.

Then again, there is a wholescale commitment to energy transition within the country, from individuals, through business and onto Government.

Many developing countries say that they do not have the means to invest away from fossil fuels. Others, such as the Trump led US, are losing the will to do so.

But things are even more complicated than that. Infrastructures are such that there could not be rapid transition anywhere in the world.

For example, we could not overnight change all diesel cars to hybrid ones. Even though the technology is there, the change is slow. In the UK, sales of hybrid and electrically powered cars are up.

But they still represent just a tiny fraction of the car market.

Risk Mitigation

This is the process by which future damage is controlled.

But it does come down to more of the same. In order to mitigate risk, it is necessary to reduce the amount of carbon dioxide we pump into the skies.

In order to do this, we need to change people's attitude to clear technologies, making them more valued than existing fossil based fuels.

Indeed, rather than mitigating, there is some evidence that renewable energy may be turning a downwards corner.

Although the overall percentage of energy in this green form is slowly increasing, closing in on ten per cent, investment is actually reducing.

So, as we reach the point where such renewables are beginning to become affordable, we are turning away from investing in them.

Over the medium to long term, if this reduction in investment continues, then the development of renewable energy will falter.

Another option which could mitigate future risk is to capture the emissions from fossil fuels and pump them underground.

This is not a long-term solution; indeed we know that storing underground could lead to long term problems, such as leakage.

But it might offer a short-term solution. The will has to be there, however, to invest in the technology to allow this.

Eighty percent of carbon emissions come from urban areas, so by working to make these greener would also help. Reducing traffic emissions and making buildings for energy efficient would both help.

But both require incentives and legislation to force people, and especially businesses, to comply.

There is an unseen complication regarding mitigation. The technology to delivering reduced carbon emissions is in its infancy. For example, Solar Radiation Management is a theory that suggests pumping Sulphur aerosols high into the atmosphere to reflect the sun's rays before then even reach the earth.

But we are a long way from achieving anything even close to this.

Yet, many countries use these distant ideas of the future to compensate for current complacency.

Supporting the Actions of Non-nation State Actors and Business

Here, the UN is looking at how entities beyond government can influence procedures to limit or reverse climate change.

These include businesses and influencers who may persuade nations to change their attitude towards such matters.

Mobilizing Finance

In this case, the UN is concerned with utilizing and coordinating finance, both private and from governments, to ensure it does the most good.

This could be developing ever greener technologies, or supporting the economies of regions to take them away from activates which offer negative impacts on the planet, such as logging.

It is encouraging that many corporations recognize their responsibilities to the planet, and nations spend parts of their tax raising capability to support solutions to climate change.

But, the UN would argue, with greater coordination, and stronger targeting, more of a difference could be made.

Deforestation

As we know from our school science, forests absorb huge quantities of carbon dioxide. Yet, since the onset of industrialization, about half the world's forests have been removed.

As parts of the world where giant tropical rainforests thrive – especially South America and Africa – develop, so the rate of deforestation increases.

If we have lost half of the planet's woodland in the last two hundred and fifty years, it

augers badly for how long the remainder
might last.

Brazil is the world leader, leagues ahead of
the nearest competition, when it comes to
chopping down its forests.

Trees are cut down to be sold as timber,
wood and fuel (ironically a double whammy
of a problem, since when the fuel burns it
releases carbons which it can, of course, no
longer re-absorb).

The land is used for farming, settlements and
to create space for mining.

Through effectively directing national
legislation, employing non-national pressure
groups and redirecting finance, three of the

UN aims, deforestation could be slowed, perhaps even halted.

The question as always is whether the will to do so exists.

Ocean Degradation

Over fishing, pollutions, the effects of climate change, and unsustainable fishing techniques such as blast fishing or using drift nets – all are combining to destroy our oceans.

Climate change's contribution is to raise sea levels and ocean temperature to the extent whereby life systems are changing.

The balances in the ocean are lost, and its ability to absorb carbon is reduced. Indeed, many scientists believe that the carbon sink provided by our oceans could be close to its limit.

Having even more carbon heading upwards to the atmosphere, when it is already failing to cope, could be catastrophic for the planet.

Although Governments talk a good game, there is little regulation of open international waters, and most nations seek to preserve their own interests.

Perhaps by taking charge of the international waters, legislating against their pollution and over fishing, the UN could do something that really helps to address the damage we are doing to our planet.

We have looked in this chapter at what the planet's most powerful body, the UN, is seeking to do to downgrade climate change. Later, we will look at how we can offer our own little bit of planet preservation.

But before this, we will consider another case of climate change in action.

Chapter Seven – Climate Change in the UK, A Case Study

In this chapter, we will take as a case study the United Kingdom and consider how climate change could impact on a specific nation.

There are a number of ways this could occur. Because the United Kingdom is too far north to experience anything by the remains of tropical storms, threats are less immediately severe than in warmer regions, but nevertheless still exist.

Agencies are preparing measures to deal with the following potential problems, as determined by climate change models

Wet Winters and Dry Summers

There will be a change in weather patterns in the UK. Of that there seems little doubt. Indeed, already we are seeing these events happening. Long spells of rain, fewer icy blasts, less snow.

These are growing characteristics of the UK's winters. Many would say, with the usual Brit's sense of poor weather, that the summers remain as bad as ever.

However, when we look at the evidence, this is not the case. In fact, there has been a gradual and significant rise in summer temperatures over the past one hundred years.

Once again, an example of climate showing the picture which short term weather tries to distort.

The average temperature has rising by about a degree in those one hundred years.

But despite occasional exceptions – 1976 is the clearest example, there was just a half degree rise until the 1980s, the other half a degree occurring in the steep increase between 1985 and 2005.

In addition, the coldest six summers of the last 100 years occurred before 1973, and the coldest third before 1983.

Of the warmest ten, six have happened in the last thirty years.

Those are the facts, and the models indicate that these extremes will continue.

Jet Stream

This is the powerful run of air that exists well above the earth's surface. It directs weather and is influenced by differences in the temperatures between the poles and the tropics.

It is the jet stream that flies directly over Britain that determines its mild climate. But should the jet stream change, which climate change would suggest should happen, there will be substantial changes in Britain's predominant weather.

Impacts

But what does this mean for the population of the United Kingdom?

Unfortunately, quite a lot. Physically, the country must prepare for floods. These will result from heavy, prolonged down pours and storm surges caused by rising sea levels and more extreme Atlantic lows.

Britain has already experienced severe flooding in parts of the West Country, London and the South East in the last five years.

They can expect more. Low pressure is exactly what it says. Less pressure is exerted on the ocean so the level rises. When that is combines with high tides, waters are forced down the narrow strip of sea to the east of the country.

The water has to go somewhere, and the low-lying land around the east coast and the Thames Estuary is especially vulnerable.

The Thames barrier was created to cope with such tidal surges, but the extent to which it will manage with the extra threats climate change brings is hard to assess.

Were it to be breached, up to two million Londoners, plus hundreds of billions of pounds worth of property would be endangered.

However, it is not just storm surges that are the problem for the UK. Heavier rainfall inevitably ends up in the river system.

As they become fuller, they spread into the natural flood plains that have been created over millions of years.

House Building

In the UK, however, much building has taken place on these natural defenses. The populations of Britain are expanding, and new housing projects are widely appearing.

Indeed, there is a plan to build over a million homes within the next five years. And these will largely be on flood plains.

Therefore, when rivers need to flood, they have nowhere to go. But they will break out somewhere. Again, the threat to life, farmland, animal habitat, property and business is substantial.

It is not all bad news, though, when it comes to climate change. As the conspiracy theorists are happy to say, the warmer winters will prevent some of the thousands of deaths each year caused by the cold.

Although, hotter drier summers will counteract this to some extent.

Also, warmer weather may help a diversification of crops and wildlife, and tourism may benefit.

But these gains could be offset by increasing droughts in the summer. The United Kingdom does not manage its water well in any case, with hosepipe bans often in force.

Droughts will damage agriculture, leading to the need for greater imports of food. Given that politically the UK is currently in turmoil, after its decision to leave the European Union, imports could well cost more, leading to financial hardship for many.

Wildlife and Plants

The United Kingdom offers a diverse habitat for numerous varieties of flora and fauna.

But, changes in climate are likely to see changes in the vegetation of the country.

That in turn will lead to changes in the animal life that exists.

Pollution

The UK health agency takes climate change very seriously. It recognizes that this is something that very much exists, and represents a serious threat to the well-being of the UK population.

One of the most serious implications it identifies relates to air pollution.

It concludes that the impact of climate change on air pollution is extremely difficult to predict, although the likelihood is that urban areas in the south and south east and conurbations in the Midlands and North West will suffer most.

It is the vulnerable who are at greatest risk from air pollution. The elderly, the very young and those suffering from other health related conditions.

Plans and money will need to be spent protecting these groups from air pollution through education, support and medical interventions.

UV Exposure

The same agency identifies the advantages and disadvantages of warmer summers. These, it is said, will lead to more time spent outdoors.

This could be good for mental health and for exercise, but there is a growing risk of exposure to increased UV levels.

This could cause an epidemic of skin cancers if the population does not become more aware of the dangers of being in the sun for too long.

Whilst in the case of the UK there are definite issues which climate change will cause, most damage and threat will result from indirect results of climate change in more vulnerable environments.

This is something that short termist climate change science opponents often fail to consider.

Whilst they might not overtly say that climate change is not their problem, failure to (for example) limit factory use of fossil fuels can only be justified in terms of benefit to their country.

But when the rest of the world suffers consequences, the global nature of industry and economics means that those impacts are felt at home.

This chapter is about the United Kingdom, and the country is far from the worst in the world when it comes to inaction.

Policies such as penalties on diesel use in cars, promotions and grants for the purchase of green transportation and help with solar energy use are just three examples of the UK doing its bit.

Nevertheless, the most significant threats regarding climate change of the near and mid future relate to impacts reverberating from the areas most effected.

Immigration

Impacts felt in the most vulnerable areas of the world, such as extreme weather conditions, deforestation and desertification (the spread of deserts) will lead to series social, economic and physical problems in the developing world.

As a result, political tensions are inevitable. And what is the result of civil war? People flee.

We can already see political problems in the Middle East leading to extensive migration.

The UK is a popular destination for political and economic migrants.

Climate change can only make this situation worse, and the UK can expect pressure to open its borders even more than at present.

Yet, already, there are deep divides in the country when it comes to immigration. One of the driving forces behind its decision to leave the EU was people's, often uninformed, views of immigration from other parts of Europe.

The UK is a very integrated society, marked by tolerance and inclusivity. But not by all, and the political situation will almost certainly worsen with increased immigration.

Equally, with the labor force needing to become increasingly skilled to find employment, more immigration will reduce the chance of the native population finding work.

This will inevitably lead to social problems.

Population Changes and Healthcare

As factors such as nutrition, medicine and lifestyle improve, so the population in the UK is aging.

Although diet issues have to some extent led to a revision of this prediction, nevertheless millennials born at the turn of the century have a significantly improved chance of living beyond a hundred.

Already, the health service is under intolerable pressure. Migration and health issues associated with pollution can only make this worse.

Food

Changing climate will lead to changing conditions in which foods thrive. Inevitably, agriculture will suffer from increased heat and drought in the summer, and flooding in the winter.

This will not only put pressure on the agricultural community, but will lead to the need for more importation of food from abroad.

Yet, the nations able to provide such food will also be struggling.

This could lead to food shortages, or more likely, significant rises in costs which will mean the poorest parts of society struggle to get enough to eat.

Political Instability

We have seen above how political instability might grow within the country, but greater risk is that instability overseas could provide a greater threat.

Migration is just one part of this. But the safety of the entire world is at risk if the divide between rich and poor grows, whilst those poorer nations have access to ever more powerful weaponry.

At the time of writing, but hopefully not by reading, the world is regarding the situation in North Korea with more than a touch of anxiety as it's errant leader Kim Jong-Un tries to stand up to the unpredictable Donald Trump.

This kind of situation will inevitably become more widespread.

But more than this, increasing poverty will lead to growth in terrorism as political and religious extremists find more and more fertile populations to exploit.

Poverty breeds problems, and climate change creates poverty.

As we can see, even for a nation as relatively free from risk as the UK, global implications of climate change will impact heavily on its future.

Chapter Eight – What Can We Do?

Things are, as we see, not looking good with regards to climate change, but there are things we can do as individuals to make a difference.

Admittedly, unless you are an oil producing sheikh, or managing director of a multinational such as Shell, one individual's actions are not going to stop the process.

But, if we all contribute a little, collectively we it will definitely make a difference.

Ease off the Gas

Up to a quarter of carbon dioxide emissions come from cars. Little steps can cut down on this production, with the added advantage of saving a little at the pumps.

Things such as ensuring tyres are fully inflated, there's no unnecessary weight in the car, and controlling acceleration and slowing down all reduce fuel usage and emissions.

Don't Overfill the Kettle

Millions of liters of water are boiled every day which are then left to go cold again.

Energy has been used to boil the water, and that energy is likely to have been created using fossil fuels.

Once again, by helping the planet, we will also be helping our bank balances.

Eat Less Meat

It is hard to believe that almost twenty per cent of greenhouse gases can be put down to the processes involved in livestock farming.

Cutting back on meat, especially red meat, is good for our health, and also saves money since it is an expensive foodstuff.

It also helps to cut down on damage to the planet.

Eat Local Produce

The carbon footprint of food from far away is considerable. The need to do this has become one of the myths of the world.

Let's take strawberries as an example. There are few fruits that are less tasty, sweet, juicy and succulent.

Those readers in the prime of life, or a tad beyond, will recall the treat that in season strawberries used to provide.

But now, for the wealthier nations at least, they are omnipresent. Yet, if we are honest with them, out of season strawberries are such a pale imitation of the 'real' thing.

They are often hard, tart and fairly tasteless. We can apply this to many foods. Buy local foods and not only are we cutting down the carbon footprint, but we are getting fresh, in season foods.

It was how we used to live, and there is a push to return to this type of food consumption. Support it, and we will start to rediscover the taste of real food.

We could even go back to growing our own vegetables. That's a thought!

Turn Off Your Computer

On average, sixty to sixty-five per cent of computer energy is used while your computer is on standby, not while it is actually doing its job.

By adjusting the settings so that it turns off when not in use, this energy is saved.

Turning down the brightness can also help to reduce energy use.

Walk More

We are all guilty of using the car for short journeys, or taking public transport when we could walk or cycle.

As well as the benefits to health and wealth, clearly there are no carbon emissions created through walking and cycling.

So, we could get fitter, more relaxed (exercise reduces stress) and save money while we are saving the planet.

Seems a no brainer really!

Let Everything Sleep at Night

While we are in the land of nod, many of us leave our houses very much awake. The television is on standby, lights are on downstairs and so forth.

But turning off our electrical equipment and lighting will make a vast reduction in the electricity we use.

Here is an astonishing fact from the UK. If every person turned their computer off at work, rather than leaving it on standby, it would be the equivalent of taking 245000 cars from the road.

Get Political

Politicians, wherever they are in the world, want votes. They also love the approbation of others, especially those who can keep them in their jobs!

Many of us are worried about climate change, but we are mostly guilty as seeing it as a problem for tomorrow.

Because it is climate about which we are talking, and climate is a long-term matter, it can seem to be something that can be put off.

Of course, we see from the Ethiopian coffee farmers and the residents of New Orleans, to give just two examples that the problem is now.

Putting the two points above, if we all contact our political leaders, demand to know their views on climate change, and what they are going to do about it, then that will lead to action.

Imagine if the Republican Congressmen who deny climate change knew that their constituencies would not be voting for them unless they opened their eyes.

Of course, it is easy for them to bury their head in the sand of big business, because they know that if they are in a secure Republican region, their futures are assured.

Pressure the Energy Companies

The power giants are reliant on customer use. If they knew that they would lose business if they were not investing in renewable power to a great enough extent, then their behaviors would change.

Cut Back on Waste

We all produce waste in our homes. Lots of it. Waste put into landfill produces methane, which is one of the most powerful and dangerous greenhouse gases.

But by composting our waste in our gardens, recycling as much as we can, we do our bit towards reducing climate change.

Shorter Showers and Turn off Taps

Showers typically pour out ten liters of water a minute. Letting the water run while we clean our teeth also generates a huge amount of waste.

So once again by saving the planet we also save money. Currently, we are paying twelve per cent of our water bills for our showers alone, plus, the cost, and emissions, of heating that water.

A lot of money to pour down the plug hole, if we think about it.

Junk the Junk Mail

Billions of pieces of junk mail are distributed each year across. But by spending a little time getting our names and addresses taken of direct mail databases, we can save that paper.

Let's face it; most of it ends up in the bin in any case. By cutting it out we are saving trees and thus helping the absorption of carbon dioxide. We are saving the energy used to create paper.

We are reducing the carbon footprint caused by transporting the rubbish. We are saving the waste created when it ends up in the trash can.

Spread the Word

Leading by example is the best way to get the message to our family. If our children see us adopting a 'greener' lifestyle, they will do the same.

In fact, children are some of the most earnest opponents of actions leading to climate change. Perhaps we should embrace their zeal.

Talking to friends and neighbors can help as well. Many of us just need a little push to take climate change as the threat to our planet, and our lives, that it really is.

How many of the New Orleans victims would still be alive had the planet not been polluted and sea levels consequentially risen?

Multiply those unnecessary deaths across the planet.

It is as though we have become trapped in a cycle of waste, of pollution, or unnecessary energy use. Although it costs us a small fortune, makes us less healthy and causes us to eat less tasty food, we allow it to happen.

Maybe now is the time for more of us to take a stand?

Chapter Nine – A Look Into the Future

Let's embark on a bit of fiction, and look at the world in the future. Our time travel machine will make three stop off points, in 2050, 2100 and 2300.

Although, it is not just fiction, while we cannot predict the future with certainty, a group of 26 leading climate scientists recently produced the Copenhagen Diagnosis, which seeks to do just that.

They based their findings on detailed mapping, rather than imagination.

Because, for every ton of carbon emissions we release now, a quarter will still be warming the planet in a thousand years.

That is a worrying statistic.

2050

Our time machine is quite sophisticated, because not only can we direct it stop when we want to, but also where we want to.

We begin our journey by visiting New York. We plan to start on Staten Island, and are glad that we remembered to fit floatation devices, because the beach we sought has gone.

Disappointed, we head south to see Brazil, arriving in Rio in time for the evening weather forecast. We tune in using our on-board computer, and close the roof.

We are in torrential rain. But, we soon discover, it is about to get worse. We find out that a month's typical rainfall will descend in the next two days.

Evacuation plans against the flooding Rio will experience are already in place. We look across to the deserted Copacabana beach, and see the remains of the shops, hotels and restaurants that used to accommodate the thousands of daily tourists.

Some still show signs of life, although their glamour is far gone. Most, though, have windows smashed or boarded up with graffitied plywood.

The magnificent edifice of the Copacabana Palace Hotel is stained, and writing has been scrawled on it frontage.

The odd light flickers, but mostly the building sits in darkness. But there is good news, because the weather forecast tells us that if we head to the north of the country, then we can escape the rain.

In fact, that whole region has escaped the rain. The longest drought in recorded history is set to continue.

Meanwhile, in the Philippines, the typhoon season has started earlier than usual. The third major storm in a month is due to make landfall within a day.

Despite claims from their beleaguered politicians that sea defenses will cope, the population knows the truth.

We set the dial forward, and in a flash, we have arrived on the east coast of the Britain. It is the year 2100.

2100

Off the coast we see armies of turbines standing still. On closer inspection, we realize that these are old machines, and are decaying fast. Towards the land, we see a few houses.

A sign announces that this is Great Yarmouth, formerly a holiday town of 40 000. It would be double that size at this time of year, because it is clearly summer, and the sun is beating down with Mediterranean heat.

Now it is little more than a wrecked village. The devastation we saw fifty years ago at Copacabana has been repeated.

We look at the rusting roller coasters in what used to be a fun fair. And we can see the supports of what we think is its pier, on which thousands would be entertained every day.

Surely a rich nation such as Britain could cope better than this? We head into London, and the Thames is much wider than it used to be.

The Houses of Parliament are gone, as has Big Ben. The cost of repair after the great storm of 2092 was too great. It was easier to start again. The breached banks and defenses would need to be rebuilt but the Thames Barrier, on the edge for years, was finally damaged beyond repair.

The good news was that the underground would soon once again be fully operational.

Flying over Europe, America, Australia and parts of Asia we are surprised to see no smoke from factories, but much evidence of renewable energy.

Perhaps the planet has finally learned? But in the tropics, we find little evidence of rainforest, just clumps here and there. We land to take a closer look.

And see a sign, written in many languages. It reads – 'Welcome to the Western African Rainforest. Here was home to mountain gorillas before their extinction.'

A map shows the extent of the remaining rainforest, Twenty square miles. But at least, we see, it is now protected.

We head to Australia and see the three great sights of the Sydney Harbor. The Bridge, the Opera House and the giant Sea Defense that stretches out to at the harbor entrance.

The great steel gates are raised at the moment. We can hear the waves crashing at them from beyond.

We stop for refreshments at a local café. There is not much on offer, and we read that their supply of coffee has run out.

But they do expect their new delivery by the end of the next month. We buy a newspaper and read of the continuing troubles on mainland Europe, as migrants fight with locals over jobs and housing.

Depressed we head forward for our final stop.

2300

It is hot outside, and we check our computer to see temperature patterns.

The world is still warming, but at a slower rate. Man has learned its lesson, it seems. We search to congratulate the world's leaders.

But whilst the work of man is still evident, buildings – many flooded - roads, factories, no matter how hard we look, all is quiet.

Fiction? We hope so.

Conclusion

Climate change is real. One day, hopefully before it is not too late, those who deny its existence, or ridicule its impact, will wake up to this.

This book will hopefully have offered more insight into the causes of climate change, things we can do to help reduce its impact and both the potential and existing consequences of doing nothing.

There have been some case studies of climate change in action, and also of how Governments are looking to prepare for the future.

We saw that the Paris Conference, in setting a two degree increase in global temperature as an acceptable limit, recognizes that containment rather than cure is the limit of our expectations at the moment.

But that some leaders have not accepted even this.

We encountered the reaction of the Trump regime. The President of the largest industrial power in the world denies the danger that climate change brings.

We hope that this book will be a call to action. There are two things we can do as individuals, and we urge you to tackle these challenges.

Firstly, cut down on your own carbon footprint, and secondly put pressure on decision makers to take responsibility for the futures of our children, our grandchildren, our wildlife, our most vulnerable people and our planet.

Fail to do so now and it could be too late in the future.

37514510R00081

Printed in Great Britain
by Amazon